滨海别墅
SEASIDE LIVING
50 REMARKABLE HOUSES

滨海别墅
SEASIDE LIVING
50 REMARKABLE HOUSES

（澳）视觉出版集团 编著

于丽红 译

广西师范大学出版社
· 桂林 ·

images Publishing

图书在版编目(CIP)数据

滨海别墅/澳大利亚视觉出版集团 编;于丽红 译.—桂林:广西师范大学出版社,2015.9
ISBN 978-7-5495-7123-9

Ⅰ.①滨… Ⅱ.①澳…②于… Ⅲ.①别墅-建筑设计-世界-现代-图集 Ⅳ.①TU241.1-64

中国版本图书馆CIP数据核字(2015)第202156号

出 品 人:刘广汉
责任编辑:肖　莉　于丽红
版式设计:吴　迪

广西师范大学出版社出版发行

(广西桂林市中华路22号　　邮政编码:541001)
(网址:http://www.bbtpress.com)

出版人:何林夏
全国新华书店经销
销售热线:021-31260822-882/883
恒美印务(广州)有限公司印刷
(广州市南沙区环市大道南路334号　邮政编码:511458)
开本:802mm×1 092mm　　1/12
印张:17 1/3　　　　　字数:30千字
2015年9月第1版　　2015年9月第1次印刷
定价:258.00元

如发现印装质量问题,影响阅读,请与印刷单位联系调换。

目录

前言	6	RUSSELL ABRAHAM
信天翁式住宅	8	BGD Architects
阿拉莫阿娜度假屋	10	Hulena Architects
自维持住宅	14	Zen Architects
拉斯阿伦纳斯海滨别墅	18	Javier Artadi
海滨别墅	20	BKK Architects
勃朗特住宅	24	Rolf Ockert Design
坎普斯湾度假屋	28	Luis Mira Architects
舒克海角住宅	30	Jackson Clements Burrows (JCB) Architects
卡皮托拉住宅	36	Park Miller
CASA ISEAMI工作室	46	Robles Arquitectos
赛德拉度假屋	48	MYCC Architecture
悬崖之屋	52	Architecture Saville Isaacs
崖顶别墅	56	T01 Architecture & Interiors
云湾度假屋	62	1 + 2 Architecture
克洛夫利别墅	64	Rolf Ockert Design
皇冠度假屋	70	Phorm Architecture + Design
柏景别墅	72	Eric Miller Architects
数字化别墅	82	Jorge Graça Costa
沙丘别墅	86	Jarmund/Vigsnæs AS Arkitekter MNAL
爱特利别墅	88	Studio 9one2
加雷别墅	92	Robert Swatt
海港别墅	96	Helliwell + Smith • Blue Sky Architecture Inc.
赫伯特住宅	98	Flesher + Foster Architects
热水湾别墅	100	Stevens Lawson Architects
奥马哈别墅	102	Daniel Marshall Architects
克里夫顿山别墅	104	Wilson & Hill Architects
克莱因瓶别墅	108	McBride Charles Ryan (MCR) Architects
灯笼度假屋	114	Eric Miller Architects
长丘住宅	122	Hammer Architects
朗埃克别墅	128	Damien Murtagh Architects
LOUTITT湾景别墅	130	Studio 101 Architects
马鲁巴别墅	134	Wright Feldhusen Architects
蛎鹬别墅	138	Ligon Flynn
板式别墅	148	David Hertz Architects
劳马蒂沙滩别墅	152	Herriot + Melhuish: Architecture Ltd
海角景观别墅	154	Andrew Simpson Architects
赛德尔别墅	158	Alexander Seidel
郊外海滩别墅	160	Luigi Rosselli
避暑别墅	162	Jarmund/Vigsnæs AS Arkitekter MNAL
冲浪海滩别墅	166	Centrum Architects
塔卡普纳别墅	170	Julian Guthrie
茶树别墅	174	Paul Morgan
里维埃拉别墅	178	Shubin + Donaldson Architects
特里格·普安住宅	182	Iredale Pedersen Hook Architects
特鲁罗住宅	188	ZeroEnergy Design
V2V别墅	190	Studio 9one2
维多利亚73号	192	Saota
KAIKOH度假别墅	196	Satoshi Okada Architects
鲸鱼海滩别墅	198	Cullen Feng
泽菲罗斯度假别墅	200	Koutsoftides Architects

前言

拉塞尔·亚伯拉罕

几百万年前，人类的祖先从丛林深处走出，并不断接近海洋，直到现在，海洋已经成为人类灵魂的一部分。今天，世界上80%的人口在海岸线60英里（100千米）范围内生活，有10亿人口依靠海洋的慷慨而获得食物。然而，生活在海边600英尺（180米）范围内的人们说他们可以尝到咸雾。心理学家现在相信仅仅是靠近大海，就能深深的影响一个人的幸福感。当我们在海边的时候，所有的感觉都能被触动。海洋不断变化的蓝光和绿光可以使焦躁的情绪变得平静。千年来，海洋对健康的重要作用已经为人们所周知，但直到现在才得以被利用。

19世纪，滨海度假区的概念在美国和欧洲就已经和铁路一起成熟了。工厂工人需要休假以便设备能够被维护保养。首先是英国，然后到了美国，工人在新开发的滨海度假区休假。后期修建的游乐场、防洪堤，以及公共浴室和酒店用来招待这些大众。更富有的人修建单独的滨海度假别墅，这些别墅是代表财富的奢华房产。从新英格兰到大西洋国家，再到迈阿密、佛罗里达，滨海小区突然出现在美国的旅游胜地。到19世纪末，英国已经有100多个滨海度假胜地。

在20世纪，铺设道路和汽车大众化的快速发展使另一层次的人们也进入滨海，并且这一现象迅速全球化。对很多都市人来说，开车到达海边的话通常只需要一小时，或者更少的时间。像加利福尼亚州南部和佛罗里达州南部的地方，海边生活已经成为人们日常的选择。在洛杉矶，滨海旅游胜地最初是电车路线的终点站。很多用于日常生活的1,100平方英尺（100平方米）的别墅已经变成了3,300平方英尺（300平方米）的迷你大厦。迷你大厦很快被细分成小部分后，廉价卖给中产阶级。这些中产阶级想在海边度过更多的时间，而不是仅仅一天。具有标志性名称的城市，像马里布和威尼斯，缓慢的并入了大城市的建筑物，但依然保留了它们自己独特的度假氛围。20世纪后期，空中旅行已经变得大众化，当初很难到达但却深深吸引人们的天堂一样的地方，现在只需要几个小时就可以到达，而不再花费几天的时间。当初那些从不列颠哥伦比亚到哥斯达黎加的沉睡渔村，已经逐渐成为了人们的度假胜地。还有那些从加利福尼亚到塔斯马尼亚岛的大牧场和乳牛场，也已经变成了新的滨海社区。全球化且正在壮大的中产阶级，拿出了部分财产，建造了天堂一样的滨海第二住宅。

对建筑设计师来说，一个滨海住宅项目可以是一把双刃剑。一方面，他们经常将传统的约束抛弃在一边。选择在滨海生活的人们，通常希望当他们在家的时候，可以感受到他们的房子能够散发出勃勃生机，更不要说他们对隐私的牺牲。一些设计在更加城市化的环境里会被禁止，但是设计师可以在海边进行这些设计。另外一方面，预算和恶劣的滨海环境能够大大的限制项目的范围和坚固性。滨海对建筑物来说，是一个恶劣的地方。长期存在的盐雾会腐蚀建筑大部分的表面，同时偶然性的飓风或者气旋可能迅速带来灾难性的损坏。对一个有灵感的客户来说，创造一个天堂一样的小别墅的挑战是不可抗拒的。从澳大利亚的维多利亚到加利福尼亚的威尼斯；从挪威的威斯特福尔到南非的开普敦，书中的设计师挑战了极限，利用各种各样的方法，建造了百花齐放的特色建筑。这些特色建筑使滨海别墅的设计上升到了更高的层次。他们结合了传统和现代的形式，在苛刻的环境下创造了非凡美丽的滨海别墅。而这些美丽的滨海别墅，无论是对业主来说，还是对前来欣赏的公众来说，都是一个令人称赞的地方。

Situated on an exclusive residential street, in Mermaid Beach, Queensland, Australia, **ALBATROSS RESIDENCE** is successful as an oasis that captures beachfront vistas and access, while maintaining a private yet expansive pool and entertaining core. The house was designed by **BGD Architects** to accommodate a growing family, multiple guests, and frequent entertaining, while the interiors were designed by Edge Design and Interiors. Entry to the residence is via a recycled timber colonnade and gatehouse that grants covered access to both wings of the house. A glazed hallway wraps the central courtyard on both the ground and first floor, allowing scenic circulation past the main amenities and providing aesthetic flow through the timber stair. Louvered glazing has been utilized throughout the home to control ventilation by natural breezes. The internal and external palette of finishes, including natural timbers and stone, creates a tropical, modern, and comfortable ambience. External finishes of recycled timber, natural stone, and copper were chosen to allow the property to further develop character over time. Tall mature trees matching the scale of the house are intentionally located about the property to frame the beachfront, main entrance, and internal courtyard. Lighting of the landscape at night creates drama within the timber battens and palm fronds, backed by the ambient aqua glow of the swimming pool. Photography: Remco Jansen

信天翁式住宅

The owners of **ARAMOANA** live on a farm in Hawkes Bay, on the east coast of New Zealand's North Island and wanted a place near the beach where family could come together on weekends and vacations. Southern Hawkes Bay is a comfortable 45-minute drive from the owner's home, but the coastal environment couldn't be more different. The hilly terrain is as dramatic as the surf beach, and according to architect Brent Hulena of **Hulena Architects**, it's quite a desolate area with only a handful of houses.

The brief was for a five-bedroom house, which could also accommodate the clients' children and grandchildren, including a large living area where all the family could congregate. With this in mind, the concrete (block), steel, and glass house is designed in an L-shape. One wing, facing the water, comprises a 49-foot (15-meter) open-plan kitchen, dining, and living area. Beyond the kitchen door are two bedrooms and a library. The shorter side of the 'L' features another two bedrooms, one of which is used as a bunkroom that doubles as a second living area for grandchildren.

阿拉莫阿娜度假屋

As the region can experience extreme winds, the architects designed two outdoor areas, one leading from the living areas to the beach on the east, the other to the west. The western courtyard is not only protected from the wind, but also benefits from views through the generous glazing on either side of the kitchen and living areas through to the beach beyond. To strengthen the link between the indoors and outdoors, all the rooms open onto outdoor areas. When it comes to alfresco dining, bifolding windows were incorporated in the kitchen, enabling easy entertaining to the eastern terrace. A fireplace in the western courtyard allows for comfortable evening dining. As the house is used year-round, Hulena felt it was important to create comfortable outdoor spaces.

While the outdoors and indoors are blurred in this home, the living spaces are pivotal to the design. 'There's nothing better than just lying on the lounges and gazing out to the beach,' Hulena believes. He also included aluminum louvers above the glass doors in the living areas, as well as over the outdoor terrace, to control sunlight. Photography: Kallan MacLeod

自维持住宅

The Jan Juc **AUTONOMOUS HOUSE** is sited on a clearing made vacant by a fallen gum tree in bushland near Bells Beach, in Victoria, Australia. The naturally occurring clearing was chosen to minimize environmental disturbance and because the space provided excellent access to sunlight in an otherwise dense tree canopy.

The client's brief and the site's remote nature called for an autonomous building that harvests its own power and water and treats its own waste on site. The resulting house, by **Zen Architects**, uses passive solar design for most of its heating, and prevailing breezes for passive cooling. The form of the house minimizes its footprint by 'weaving' through the trees. The main entry hallway follows the path of a well-worn wallaby track that cascades down toward a seasonal creek. The darkness of the entry provides a dramatic contrast to the bright, north-facing, habitable rooms, giving the sensation of walking into a sunny clearing.

Recycled wharf posts are used as structural framing, continuing the rhythm of tree trunks in the dense forest outside the house. They also provide a historic and coastal context as they were originally part of the old Geelong pier, nearby. Radially sawn weatherboards are cut to minimize waste and detailed to resist bushfire attack. Photography: Sharyn Cairns

Las Arenas is a gated community on the Peruvian coast, 62 miles (100 kilometers) to the south of Lima. The surrounding desert landscape is inhospitable and provides limited context for new buildings. Uniform building regulations usually imposed by the governing bodies of such exclusive communities also leave little room for individual expression. Fortunately, architect **Javier Artadi** was able to bend the rules to create a new typology for beach houses on this desert coast.

The 2,300-square-foot (700-square-meter) **BEACH HOUSE IN LAS ARENAS** was designed as a vacation home for a couple with three children. Conceived as a series of box-like containers, the house is designed for flexible use and the seamless integration of spaces, expanding the conventional, simple beach house program.

Weightlessness and freedom are important characteristics, achieved by the use of cantilevers, folded planes, framed views, suspension, and deep recesses. Sunlight control has been achieved by the strategic 'slicing' of the building volume to produce shady outdoor rooms and indoor comfort. The living-dining room is integrated with the terrace and swimming pool, and the bedroom and services areas surround the main volume. The kitchen and the main bedroom are both connected visually with the terrace and swimming pool and, by extension, with the horizon. Photography: Jacqui Blanchard

拉斯阿伦纳斯海滨别墅

BEACHED HOUSE continues **BKK Architects'** interest in the curation of the domestic as a sequence of unfolding spaces, which deny then release views. The journey throughout the house is undertaken via a series of subtly shifting spaces that alter one's orientation to both climate and terrain. BEACHED HOUSE was conceived formally as an exercise in volumetric origami—the folding of spaces over and upon each other. In this way, the house resembles a small village or informal site occupation that has aggregated over time. There are a number of these folded spatial sequences within the house that allow for playful discovery and encounter, as well as opportunities for varying connections between spaces. The home offers the owners a range of readings and differing options for occupation. It is intended that living in this house will be an unfolding series of moments, linked closely to climate and site, which will continually delight and surprise. Photography: Peter Bennetts

海滨别墅

勃朗特住宅

The client approached **Rolf Ockert Design** to create the house of their dreams on a site perched high over the Pacific Ocean—a home that would make them feel like they were on vacation every day. While the view was fantastic, this site in Sydney, Australia, was very small and suffocated by overbearing neighboring dwellings. The finished house feels generous, as if it is alone with the ocean and the sky. High side walls offer a sense of privacy but also provide mass for a comfortable indoor climate, with continuous highlight windows for the enjoyment of 360-degree views of the sky. The large face concrete wall dominating the space has slim slot windows, allowing teasing glimpses of the ocean when entering the house, while effectively cutting out the visual presence of neighboring properties.

BRONTE HOUSE opens itself up completely to the east, which presents stunning water views. This also allows the house to capture the constant ocean breezes to cool down throughout the year. These breezes are easily regulated by a plethora of ventilation options from sliding doors to operable louvers.

A rich but reduced palette of strong earthy materials—from concrete to timber flooring and ceilings, rust metal finishes, and thick textured renders—contrasts with the fine detailing of the interior and anchors the residence against the airy, light aspect created by the opening to the views. Photography: Rolf Ockert Design

坎普斯湾度假屋

CAMPS BAY HOUSE is a vacation home for an individual who often invites guests to stay. There was a need for a dual-purpose space that could be used as an open-plan studio for the client, or transform into more singular private spaces for visitors. The site's privileged position and Cape Town's agreeable climate informed **Luis Mira Architects**' design process in which the brief also required all rooms to have sea views. The house focuses on the 'geographical room' of Cape Town's Camps Bay and the Atlantic Ocean, while the peaks of Lion's Head and Table Mountain form the backdrop. Design centers on framing views toward the sea in each room, and opening up terraces toward the mountains behind. CAMPS BAY HOUSE invites a subtle journey throughout, moving through open spaces and the interior, where continual glimpses of the building and surrounding landscape merge without ever revealing the entire residence in a single instant.

On the ground floor, two glass-walled courtyards enjoy sea views. They provide rooms to the front of the house with proximity to the sea and circulation, fresh air, and light to the back of the house. One courtyard is built around a corridor and connects the bedrooms, while the other is integrated inside the main bedroom as part of the ensuite area.

A desire to bring the outdoors inside was achieved by contrasting the exuberant surrounding landscape with the 'blank canvas' of the interior, through the use of neutral and natural materials. While 'peeling off' the volume of each interior space to the north—the source of sunlight in the southern hemisphere—enhances the sense of inviting the outdoors in. CAMPS BAY HOUSE's ultimate luxury is the constant extending and opening of its inside spaces to embrace the unique and exquisite South African landscape and climate. Photography: Wieland Gleich and Mira Architects

The owners of **CAPE SCHANCK HOUSE** took a leap of faith in accepting the initial sketches presented to them, as the house resembles a 'burnt-out log', with a 29.5-foot-long (9-meter-long) cantilevered wing. The form of the house evokes a log, or two branches, when both wings of the house are seen in profile. **Jackson Clements Burrows (JCB) Architects** considered several aspects of the property, located at Cape Schanck, on Victoria's Mornington Peninsula in Australia. One of the strongest images was of an old campfire on the property, left with a burnt log.

Architect and Director Jon Clements says they also wanted to capture the entire vista over Port Phillip Bay. Using one of the covered dunes as an anchoring point, the house extends in two directions. One wing comprises the kitchen and living areas, while the other, with the main bedroom, ensuite and study, is perpendicular to these spaces. Supporting these two wings is a plinth-like base, containing two guest bedrooms, a bathroom, and second living area. 'The owners were fairly open when it came to the brief. They didn't mind the log analogy. But they wanted the "log" to rest at one level,' explains Clements. 'When they, [a semi-retired couple], come down on their own, there's no need for them to go downstairs,' he adds.

JCB could have designed an all-concrete house, given the area's predisposition for bushfires. However, with a 164-foot (50-meter) clearance around the house, the architects were able to explore the use of timber. Fire-hardy timbers such as spotted gum were used to clad the steel and timber-framed house. The timber was stained black and picks up the trunks of nearby tea-trees, while cedar was used for the windows and sliding glass doors.

舒克海角住宅

While the house appears relatively exposed, the front door is at the end of a winding driveway at the top of the hill, discreetly placed next to the garage. Past the front door, the expansive living areas take in dramatic views over Bass Strait, Point Nepean, and Port Phillip Bay. The only distraction to the main vista is the large deck and swimming pool, adjacent to the kitchen and dining areas. The palette of materials used in the kitchen and living areas shows different aspects of the hollowed form. The graphic black and white kitchen was treated like a cube, with joinery appearing to be carved from a solid form. In contrast, the lounge area features warmer materials such as spotted gum around the fireplace and a cedar-lined wall. 'The cedar has a warm feel about it. In contrast to the blackened spotted gum it's like the protected heartwood that hasn't been burnt,' says Clements. Photography: John Gollings

卡皮托拉住宅

The Monterey Bay stretches in semi-circular fashion for about 62 miles (100 kilometers). The region was originally developed as a logging and commercial fishing center, but later transitioned into a string of seaside resorts in the early 20th century. Capitola sits at the northern edge of this crescent bay, a modest sister city to its more robust urban sibling, Santa Cruz. Capitola's pastel beachfront cottages, modest ocean-side eateries and gift shops are a quaint throw back to simpler times. Monterey Bay is warmer and calmer than the Pacific Ocean next to San Francisco, and as a result, it is a draw for surf-seekers from the San Francisco Bay Area, 93 miles (150 kilometers) to the north.

The family that commissioned architect **Park Miller** to design the **CAPITOLA HOUSE** had vacationed in the area for years. It was an easy drive from their home in the Bay Area and a great place to spend a weekend with family and friends. The family had owned a nearby house for some time, but when a beachfront cottage was put on the market, they jumped at the opportunity. The existing house sat on a spectacular hillside site overlooking the beach and ocean, but was in terrible shape. The new owners decided to tear down the cottage and start fresh. Town planning required that any new building fit within the existing footprint and conform to the hillside character. The owners asked Miller to design a house that met the design guidelines and blended in with its eclectic neighbors. Since they enjoyed entertaining and had teenaged children, they wanted a house that could accommodate them and the inevitable numerous weekend guests.

Park Miller's solution was to excavate into the hillside to create living space below grade. He gave the house a reverse floor plan using the upper floor for public functions and the lower floors for private. The bottom floor has a guest bedroom and kids' playroom. The second floor contains the main bedrooms and the top floor, the living and dining great room.

The owner's wife had a fondness for all things New England, and did her best to make this a Yankee retreat parked in the middle of the California coastline. Her palette and furniture choices strongly emote Cape Cod with a dash of whimsy added in for fun, such as a newel post shaped like a light house or a bar top of glass with a treasure chest of faux pirate booty and seashells on display just underneath. She also spent countless hours collecting historic coastal memorabilia, which she used to decorate the house and create something of a local historical museum.

Working within the tight building codes and restrictions, architect Park Miller and builder Paul Mehus managed to hide a big house on a very small lot and have it look like it was always there. Photography: Russell Abraham

Spectacular coastal views characterize **CASA ISEAMI**—the first physical manifestation of the ISEAMI Institute (Institute for Sustainability, Ecology, Art, Mind, and Investigation), in Playa Carate, gateway to the Corcovado National Park, Costa Rica. The house is the institute's main multifunctional area for hosting activities including research, meditation, training, yoga and relaxation, specifically on the first-floor terrace. **Robles Arquitectos**' design process evaluated ten principal elements: site, climate, energy, water, materials, environment, atmosphere, cost, and innovation with the use of passive strategies and implemented processes. These elements were analyzed to develop a design and management plan during the building lifecycle, which would reduce the house's potentially negative impact on its pristine natural environment, and conversely minimize the negative effect nature could have on the building. In short, the architect's main objective was to create an extremely low-maintenance house.

The secluded project site is located approximately 18 miles (30 kilometers) from the nearest town, Puerto Jiménez, and the house does not have mains electricity or water. The institute had to invest in a house that is entirely self-sufficient for its energy and water needs. The structural and electromechanic design was inspired by the exoskeleton of an insect, which created open spaces between walls and ceilings and eliminated enclosed space. Resulting in beneficial indoor air quality and avoiding conditions for mould and insect infestation which is a common problem for projects in tropical climates. Passive design strategies have been successfully implemented to handle sun exposure, relative humidity, natural illumination and ventilation, according to bioclimatic circumstances.
Photography: Sergio Pucci

CASA ISEAMI工作室

赛德拉度假屋

Located in Cedeira, a small tourist and fishing town in the northwest of Spain, this vacation home consists of six modules, approximately 20-feet-long (6-meters-long) and 10-feet-wide (3-meters-wide). **CEDEIRA HOUSE** was designed by **MYCC Architecture**, manufactured in three months and constructed in three days, involving little environmental disturbance to its beautiful site on the northeast corner of the Iberian Peninsula.

The tree silhouettes cut into the house's Corten steel cladding not only provide an interesting natural daylight display, but blend harmoniously with the slender eucalypt forest in the background and the surrounding harvest fields. Corten itself was chosen because of its relationship to the traditions of the region's fishing villages as it was used for the construction of boat hulls; while its patina and changing color creates a lively image that relates to the natural environment.

The increasing demand for sustainable homes, or at least homes with elements of sustainability, has seen a resurgence in the popularity of prefabricated, or modular homes, once disparaged as cheap, mass-produced, carbon-copy kit homes. But this image has undergone significant re-evaluation in recent years, not least because of interesting designs like CEDEIRA HOUSE. Photography: Fernando Guerra, FG+SG

悬崖之屋 Located on a clifftop south of Sydney, Australia, **CLIFF HOUSE** enjoys expansive views of the Pacific Ocean. **Architecture Saville Isaacs**' aim was to create a peaceful environment for the clients on a challenging but visually stunning site. A place in which everyday living is a joy and every simple function would give cause to pause and reflect, by connecting each function to surrounding nature.

CLIFF HOUSE's layout and design is suitable for the needs of an older couple without children. Single car parking has been provided as the owners use the nearby railway to commute to the city. At the same time, it incorporates ecologically sustainable development principles and sustainable, recycled, low or non-toxic, and cost-effective materials.

The design responds to the challenges of the south-east orientation, with windows strategically positioned to maximize winter sunlight penetration. Rainwater is captured for toilets, laundry, and irrigation. Blackwater recycling was extensively investigated, but pump-out was installed due to unstable geotechnical site conditions, which prevented water dispersal on site.

CLIFF HOUSE relies on sea breezes for cooling and a heat-pump hydronic system augmented by wood fires for space heating, as there is no gas supply. The built outcome was the result of an extensive process, including the builder and structural, hydraulic and geotechnical engineers, to resolve the significant challenges posed by the unstable cliff-edge site. The cantilevered architectural forms enable footings to be kept away from the eroding sides of the site.

Cost-effective, readily available, and locally produced materials have been used in innovative ways with the materials' raw beauty providing the sense of luxury and selected to ensure minimal impact on the local environment, the greater environment, and the inhabitants' health. Materials used are standard components chosen to minimize embodied energy and cost. The building is essentially galvanized steel, plasterboard, fiber cement sheeting, recycled timber, and laminate joinery. How these materials are put together and expressed creates the quality of CLIFF HOUSE's space and architecture. Photography: Kata Bayer @ Kata Bayer Photography and Architecture Saville Isaacs

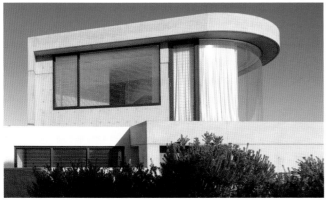

CLIFFBROOK HOUSE was a dilapidated and dysfunctional house on a superb clifftop site overlooking Gordons Bay, immediately south of Clovelly in Sydney, Australia. After acquiring the site, the client engaged architects **T01 Architecture & Interiors** to redesign the house, capitalizing on its position. As it was also to be a functional family home, the residence needed to be more than just a viewing platform. The initial client design brief referenced the concrete bunker, a defensive military fortification that would sit comfortably camouflaged within the cliff face and surrounding environment.

崖顶别墅

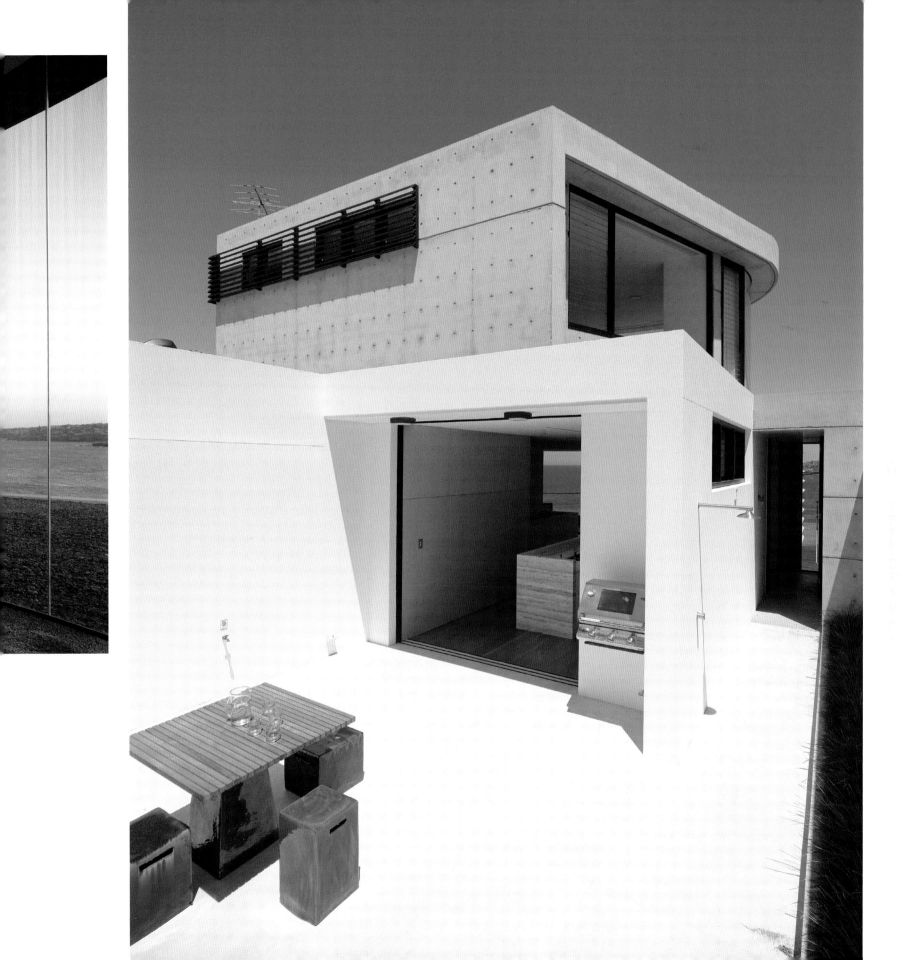

The project became an experiment in form and materiality, embracing the overt beauty of the location, the spectacular views out over the Pacific Ocean, and the site's other more subtle charms; the smell of sea salt, and the colors and texture of the surrounding sandstone cliffs. The raw concrete entry, intended to evoke the unpolished, eroded nature of the site, is on the middle level and leads into a large, open-plan living, dining and kitchen area that serves as the family's primary space. Below, on the lower level, are the children's bedrooms and a media room. The upper story houses a master suite that looks to the southwest, over the Pacific. Photography: Brett Boardman

云湾度假屋

Located at the southern end of Bruny Island in Tasmania, Australia, **CLOUDY BAY HOUSE** has three bedrooms and was designed as a vacation home for its American owners. The 82-acre (33-hectare) beachfront site comprises a combination of farmland and heath-covered coastal dune. A primary objective for the architect was to create an appropriate intervention that minimized impact on native flora and the natural landscape, while regenerating the landscape by removing weeds and pine tree remnants, and reintroducing native species. **1 + 2 Architecture** has balanced the conflicting objectives of achieving vistas to the landscape and ocean beyond, while minimizing the imposition of the house in views back to the site from key public external view points. Conceptually, the house comprises two primary zones interconnected in a conventional 'bi-nuclear' plan. The two zones are conceived as an inversion of each other in terms of function and form—an open-plan, lightweight, transparent structure forms the living zone and a heavier, enclosed, protective structure organizes the sleeping zone. The two areas straddle a series of timber and concrete external decks strategically positioned in relation to adjacent internal areas, sun, specific views, and shelter from prevailing winds. Materials were chosen for their environmental qualities, fitness for purpose, and appropriateness to the site and its context. An example is the use of macrocarpa cladding. This timber, harvested from redundant farm windbreaks, has excellent durability and has been left untreated, allowing a natural patina to develop. Photography: Jonathan Wherrett

CLOVELLY HOUSE sits on a spectacular seafront location in Sydney, Australia. **Rolf Ockert Design** constructed the house around a central void with a row of large operable skylights above it, angled to the north. While the site already included wonderful views of the cliffs across the bay, a slight swing of the external wall suddenly created expansive open water views across the Pacific Ocean, in addition to the more intimate bay views.

克洛夫利别墅

The unusual but elegant roof shape invites sunlight in while still allowing neighboring residents to enjoy water views over the lower end. The expressive angled concrete wall mirrors the negative roof shape, resulting in complex façade geometry along the main face, enhanced by the movement of ever-changing shadows over the shapes. The site is prone to relentless cold southerly winds so the house is designed without any opening windows facing south. Instead, large, frameless, floor-to-ceiling double-glazed elements retain an intimate visual connection with the ocean.

The light void also contains the central circulation, the stairs. There are two levels to the north of the stairs and three to the south, taking advantage of the natural slope of the site. The main living space is on the entry level, connecting it with the northern garden and pool, providing a generous central family area. Upstairs are the bedrooms, while the southern lower level includes several areas for more individual activities, such as the study, studios, and library.

The materials for the house were chosen for suitability from the harsh corrosive salt-spray environment and to represent the location—reminiscent of flotsam, rich but weathered. The natural palette of coastal colors—the greys and red found in the rocks, the blues and greens of the ocean and sky— provide the canvas of hues and textures that dominate large parts of the house. The resulting material palette relies on very few elements: strong raw concrete; a dark zinc roof which is allowed to weather; dark timber floors and joinery offset with white walls and joinery faces, along with the ubiquitous glass.

Operable skylights and floor vents all but eliminate the need for air conditioning. High-performance insulation and double-glazing throughout, in combination with the high thermal mass in the house, allow for the utilization and storage of the northern solar heat gain in winter, keeping the house warm during the colder months.
Photography: courtesy Rolf Ockert Design

皇冠度假屋

CROWNEST HOUSE is perched above the township of Agnes Waters and 1770 on Queensland's Discovery Coast. It commands a view to Rocky Point and the last surf beach on the eastern coastline of Australia before the surf surrenders to the Great Barrier Reef. **Phorm Architecture + Design** designed it as a retreat for a young couple and as a place to host friends and family.

The house coils around itself on the top of the hill, with the bedroom tower anchoring onto a rocky outcrop and living quarters cantilevered off the hillside. The plan responds to intense mapping and interlocution with the broader site. The form creates a central private deck that addresses the principal view and panoramic sweep. Open to the night sky, the deck relies on 'shadow-casting' from the deck awning and the massing of the house to shade and protect. The east-facing, timber-screened wall modulates the morning light and becomes a lit canvas at day's end. The timbers will be allowed to age to a silver grey with the passing of time.

The interiors form a chain of relaxed coastal volumes, tempered from the intense tropical light. Expressed detailing in local hardwoods and timber veneers contrast with articulated white plaster and polished concrete. Movement through the interior of the house is choreographed to reveal selected views of landscape. The clients have noted that the interiors are akin to 'living inside a musical instrument,' inspiring raised voice and song.

Journeying through the house one finally arrives at the 'crow's nest.' Like a pulpit addressing the sea, the cantilevered white cube projects beyond the edge of the house, leaving one to hover in the breeze, looking out to the blue horizon.

Photography: Keith Burt

柏景别墅

Carmel by the Sea is an artist community and beach town located on California's Monterey Peninsula, a region boasting some of the State's most beautiful coastline. As an early twentieth century rendezvous location for west coast literati, the town became both a magnet for creative souls as well as a weekend retreat for the more affluent San Franciscans.

A family with young children wanted just such a retreat next to Carmel's exquisite beach. The house they chose was a small, post-war creation in poor condition and too small for their large family. The house also had some tight deed restrictions on lot coverage and building height. They hired **Eric Miller Architects**, a well-known regional architect with experience in both classic and modern styles, to create a house suitable for their family and buildable for the site.

Miller's solution was to lower **CYPRESS VIEW HOUSE** to grade on the ocean side and cut into the slope on the rear so that the bedrooms were slightly below grade. By giving the house a lower story, he found space for a guest bedroom. Working around a very mature cypress tree anchored in the lot's corner, Miller oriented the public living areas toward the beach and ocean. Miller designed the great room with a 12-foot (3.7-meter) ceiling, thereby maximizing its volume and drama while staying within the height restriction. The owners were fond of the New England shingle-style beach houses they remembered from their childhood. Miller obliged with a Nantucket-style house design using cedar shingles, white trim and classic details. Both the family and dining rooms open onto a small stone patio with easy, unscreened access to a city street and the adjacent beach. During daylight hours there is constant foot traffic of beachgoers, surfers, joggers and dog walkers passing by the house. Miller's design welcomes the beachgoers as part of the human fabric of living in a small beachfront community. As one resident told me, 'You have to learn to share this magnificent place with everyone.' Photography: Russell Abraham

This relatively compact house is sited on a hill overlooking the Atlantic Ocean in Oeiras, Portugal. **DT HOUSE** has been adapted to make the most of the Oeiras climate, with its mild winters and warm to hot summers. Designed with intelligence and foresight, this minimalist house blends perfectly with its surroundings, and demonstrates the effectiveness of simple passive strategies.

Architect **Jorge Graça Costa** salvaged material from the demolition of the pre-existing home for the foundations of the new house. Material was also recycled for use in the concrete pathway and outdoor courtyard. All non-recycled materials used during construction were chosen based on their durability and potential for future reuse and recycling. For example, cork has been used for floor insulation in the upper level.

数字化别墅

On the south-facing side, the house consists of two levels of floor-to-ceiling double-glazed windows that harness and retain heat from the winter sun, while simultaneously providing plenty of natural light. The design also prevents solar gain during the summer.

Beyond its typical uses and the stunning visual element it brings to the house, the strategically positioned pool is designed to act as a natural evaporative cooling system. The air cooled by the evaporation of the water is moved through the house via southerly breezes from the Atlantic, circulating air through the two levels of the house. This completely natural process is so effective and efficient that the house uses four times less energy than a typical Portuguese house for cooling during summer. The water can also be used for irrigation purposes as it is chemical-free. Photography: FG+SG

DUNE HOUSE is located in Thorpeness on the Suffolk coast, in the east of England, and replaced an existing building on the site. Designed by **Jarmund/Vigsnæs AS Architekter MNAL**, the house is a vacation home for rental and was commissioned by Living Architecture, a social enterprise founded to revolutionize architecture in the British vacation rental sector. For planning permission approval, it was important to relate to the existing typical British seaside strip of houses. The roofscape—the bedroom floor—somehow plays with the formal presence of these buildings, and also brings to mind a romantic memory of vacations at bed-and-breakfasts while travelling through Britain.

沙丘别墅

The ground floor contrasts this by its lack of relationship to the architecture of the top floor. The architectural ambiguity of the house also addresses the programmatic difference between the private upper floor and the social ground floor. The living area and the terraces are set into the dunes in order to protect them from strong winds, and open equally in all directions to allow for wide sweeping views of the coastal area and sea. The corners can be opened by sliding doors, which emphasises the floating appearance of the top floor. While the materiality of the ground floor—concrete, glass, and aluminum—relates to the masses of the ground, the upper floor is a construction made from solid wood, and the cladding is stained dark to match the existing gables and sheds found in the local area. Photography: Nils Petter Dale

The client for **ETTLEY HOUSE** is a well-known commercial glass fabricator. Having worked with many of Los Angeles' best architectural firms, he had an intimate knowledge of their skills and capabilities. He chose **Studio 9one2** for both its design prowess and masterbuilder skills. Ettley was looking for an architect who could be creative with big glass and Patrick Killen was that person.

Sitting only a few blocks from the ocean, this tight corner lot asked for something dramatic. The client wanted Studio 9one2 to feature his company's commercial glass skills. Killen obliged with a house that is a study in solid void relationships and a 'glass showcase.' 14-foot-high (4-meter-high) glass bays pop out from a commercial window wall façade and are nestled under a huge aluminum and wood canopy roof. At certain times of the day, the floor-to-ceiling glass bays reflect the blue sky on the deeply tinted blue glass and create the illusion of blocks of frozen blue Pacific Ocean. Interspersed with the glazed bays are slatted mahogany bays that create a solid offset for the dark blue glass, creating the appearance of a modern seaside sculpture.

爱特利别墅

The ascending lot achieves commanding views of the water, while providing a cityscape foreground to the setting. The master suite, located to the front at mid-level, has a glass floor sitting area, which overlooks a reflecting pond and garden below. This patio is located off the theater/family room. On the side of this master suite bay, Killen designed a bamboo garden framed by a vertical mahogany trellis. The bamboo has grown up through the building trellis and provides needed privacy to the master suite and the main living spaces at the top level. This vertical garden gives a Zen-like feel to this corner of the house.

Studio 9one2 is known for its meticulous staircase designs. The majority of beach houses in the Los Angeles area are designed as inverted plans with the living spaces at the top level, in order to achieve the best ocean views. As a consequence of this design rule, stairs are in frequent use by guests and occupants alike. With this in mind, Killen decided to lift a page from M.C. Escher and create a staircase reminiscent of his famous etching. He accomplished this by having mahogany steps 'march up' to a square glass landing and then continue on the next side of the square. The effect is uncanny. All that was missing was the upside-down scale figure on the top stair. As in many other Studio 9one2 projects, the outside treatments are continually being brought inside. The roof soffit has become the top floor's ceiling and is exposed as wall panelling in rooms at the rear of the house. Terrazzo, mahogany, and glass are alternately used as interior flooring. The open floor plan with its giant glass bays, soaks almost every corner of the house with intense daylight. The ETTLEY HOUSE is not for the private or the faint of heart. But in a region where being public is very much a part of being, it is a work of architecture destined to become a landmark.

Photography: Courtesy of the Studio 9one2

加雷别墅

Robert Swatt must have had the architect Richard Neutra in the back of his mind when he designed the **GARAY HOUSE**. Set into a steep hillside overlooking San Francisco Bay, its sleek tan lines stand out against the verdant and rock-studded hills. The travertine-clad piano nobile flows seamlessly past retracting walls of glass onto agave-bordered patios and spectacular views of the bay.

A narrow lap pool is like an exclamation point as it terminates at a wall of Jerusalem flagstone and the main living space. A steel moment frame is hidden carefully behind metal cladding allowing for unobstructed fenestration and 90-degree glass corners. Roof overhangs that stretch horizontally in three directions control solar heat gain and provide covered patios and decks for year-round outdoor living. The clear span glass yields ever-changing day and night vistas of the bay, the Golden Gate Bridge, and San Francisco, making this Marin County hillside one of the most desirable places to live in Northern California.

The Garays lived on this site for a number of years before they decided to do a major remodel. After several false starts, their modest remodel became a completely new Swatt-designed structure that expanded the footprint of the original house significantly. Strict zoning mandated that the house have a low profile on its prominent hillside site. Swatt used a handful of architectural tricks to counteract the city's height restrictions and give the interiors a bright and open feel. Pop-up clerestories and skylights bring in light from four directions and offset the intense light from the bayside window walls. A tight entry hall steps down into an expansive piano nobile with only a subtle change in ceiling height. Meticulous detailing, elegant materials, and a soft palette of earthen hues give the house a mid-20th century Southern Californian look.

Photography: Russell Abraham

海港别墅

Given an almost impossible but spectacular site—a solid granite cliff overlooking Eagle Harbour, West Vancouver, British Columbia—**HARBOUR HOUSE** was formed as a continuation of the natural topographic lines, resulting in a series of dynamic curves and flying roofs sweeping across the rockface. Particular attention was paid by **Helliwell + Smith • Blue Sky Architecture Inc.** to the low-rise curving roof forms to help reduce the visual scale of the house, minimize the overall building height, and help open the house to natural light and views over Eagle Harbour. The split-level planning helps accommodate the vertical circulation throughout the home. The house form steps back from the road as it rises up the cliffside.

The palette of materials chosen for HARBOUR HOUSE is natural, durable and—as much as feasible—local. Exterior walls and fascias were constructed from clear natural red cedar siding, fiber cement siding, and glass. The retaining walls for the driveway, entrance stairs, and the foundation are made from granite at the site and architectural concrete.

HARBOUR HOUSE's partially covered terrace allows for outdoor living in all weather conditions, while large overhangs protect the building from rain and sun. The home has a nautical feel in some of the detailing, and the curving cedar walls recall wooden boat hulls. Its design builds on the tradition of west coast modernism, creating a living environment inspired by and appreciative of the dramatic landscape of which it is a part. Photography: Gillean Proctor

Tearing down a 3,229-square-foot (300-square-meter) cottage and building a new 9,149-square-foot (850-square-meter) house on roughly the same footprint can be a daunting task, especially on a lot facing the ocean. According to architect Bill Foster, of **Flesher + Foster Architects**, there was strong opposition to demolishing the existing cottage, but ironically, almost none to its replacement. The solution was a compact, two-story structure. Main living spaces are on the second floor, and bedrooms and parking on the first floor. Living and dining spaces enjoy wraparound ocean views, while more pedestrian rooms such as the kitchen, bathroom, and laundry are positioned to the rear of the building. A small but cozy patio garden, with an outdoor fireplace and a privacy wall, was constructed along the alley side of the building, providing a sheltered external living space. This compact house makes a definitive and uplifting statement on a street filled with eccentric and eclectic oceanfront homes. It neither fights with nor blends into the shoreline's residential scale. **HEBERT RESIDENCE** is empirical proof that modernism can exist happily in a small-scale residential environment.
Photography: Russell Abraham

赫伯特住宅

HOT WATER BEACH HOUSE is located at one of New Zealand's most popular beaches—Hot Water Beach, derived from the hot springs that filter up through the sand. This house was previously part of a caravan park, whose owner decided to subdivide the beachfront property and build a home for himself, his wife and their grown-up children. Designed by **Stevens Lawson Architects** on one of the largest parcels, approximately 21,528 square feet (2,000 square meters) in area, 'This is their permanent home, but they wanted an informal design, like a beach house,' says architect Nicholas Stevens, who worked closely with fellow director, Gary Lawson.

The single-story house is fully clad in cedar stained black. The only elements not stained black are the alcoves of the verandas, which are stained dark brown. 'It's similar to biting into an apple and seeing a different color. But the brown also provides a warmer color as you enter the different wings,' says Stevens.

HOT WATER BEACH HOUSE is divided into two wings. One wing includes the kitchen, dining and living areas, with a dark stained timber battened screen separating the dining area from the living area. This wing also includes a study, as well as a rumpus room, which has both a television and pool table. The second wing is for the children. As they are grown, the owner wanted a separate wing for them, which comprises three bedrooms and bathrooms. Linking the two pavilions is a glazed breezeway with colored glass panels either side. 'The color creates magical effects at various times of the day with the sunlight coming through,' says Stevens. The glazed link also provides protection for an internal courtyard.

Stevens Lawson Architects regularly design black homes. 'Black houses are extremely popular in rural New Zealand, as well as along the coast. There's a tradition of using creosote (a bitumen product),' says Stevens. This particular house called out for a black treatment, being on the edge of a pristine beach. 'It's an extremely sensitive site, so we wanted the house to blend in with the trees and appear recessive to the beach,' adds Stevens. 'The house is low slung so it doesn't affect the environment.' Photography: Mark Smith

热水湾别墅

Daniel Marshall Architects in New Zealand didn't meet the owners until the house was being constructed. At the time, the owners were living in the United Kingdom, planning to move back to New Zealand permanently. The design of the house evolved by phone and online, while the final stages relied on a DVD with an animated film, together with a physical model of the house, shipped over to the United Kingdom.

The initial brief for the **HOUSE AT OMAHA**, an hour's drive north of Auckland, was Cape Cod-style. 'This is not a style the architects consider themselves known for, as they prefer to create contemporary homes,' according to Daniel Marshall. To find a compatible path, he suggested the owners look at a book on the architecture of the Hamptons (Long Island, New York) in which many of the homes are quite contemporary.

奥马哈别墅

The site, perched above the beach, was also influential in the design. As it is relatively exposed to southeasterly winds, the architects felt a protected courtyard space was required. Consequently, the cedar-clad and glass house features a courtyard garden, protected from the wind by a single-story living pavilion. 'We wanted the owners to be able to look though the living areas to the sea, rather than feeling closed off ... you can see the water from wherever you are in the house,' explains Marshall.

On the ground floor are the kitchen, dining and living areas, together with a second, raised sitting area. The two living areas are separated by American oak joinery, one side functioning as storage, while the other side (the sitting area) includes an open fireplace. At the front of the house, facing the street, are three bedrooms and a bathroom. 'On the first floor are the main bedroom and ensuite, together with a dressing area. There is also a separate bunkroom. Marshall included a balcony, accessed via the main bedroom, to allow the sea views to be enjoyed at all times. 'It's a reasonably transparent house. The coastal dunes are integral to the design,' says Marshall, who included glass sliding doors in most rooms.

While the house isn't Cape Cod-style, there are finely sculptured nooks within the essentially rectilinear form. The cedar plywood ceiling in the sitting area is faceted. And three smaller canopies protruding over the deck have a fine sense of craftsmanship. 'It is contemporary, but it's not just a minimal glass box,' says Marshall. Photography: Daniel Marshall

克里夫顿山别墅

Wilson & Hill Architects was commissioned to undertake this residence on a clifftop site in Christchurch that overlooks Sumner Beach and the mouth of the Avon Heathcote Estuary, with panoramic views from Scarborough Heads around to the Southern Alps. The client's brief described the spaces required and the 'feel' of the house for those living there and for guests. The brief stated the house should be calm with beautiful views—sophisticated yet relaxed.

HOUSE—CLIFTON HILL is built over three levels; the top level is the entry, the middle level contains garaging and bedrooms, and the lower (ground) level accommodates the living spaces and a guest bedroom. The house's orientation and the rooms' positioning ensures that each living space and bedroom has a view out over Pegasus Bay or Christchurch. The stone panels that clad the house give it a feeling of solidity, permanence, and a strong connection with the ground, with living spaces opening directly onto stone terraces and flat lawns.

One of the challenging aspects of the design was how to bring people from the street at the rear of the site to the cliff edge at the front of the site so they experience the full effect of its spectacular views. This has been achieved through the design of a journey starting with a level 'bridge' crossing the gap between the street edge and the house's entrance. Once inside the house, visitors descend down the site through a three-level gallery space to the living areas stretched out along the clifftop. A timber deck provides the final step out to the cliff edge.

The gallery provides the circulation and ordering spine for the range of formal and informal spaces that connect to it and fulfills the client's request for a space to display his extensive art collection. The bedrooms, living areas, and swimming pool with pool house are placed at right angles to the gallery, creating sheltered lawned courtyards and terraces on either side of the house. The master bedroom balcony offers dramatic views directly down to the base of the cliff some 50 meters (165 feet) below. Photography: Stephen Goodenough

The owners of **KLEIN BOTTLE HOUSE** were after something completely different from their city home, a traditional Californian Bungalow. Located in Rye, on Victoria's Mornington Peninsula in Australia, the house, designed by **McBride Charles Ryan (MCR) Architects,** couldn't be more dissimilar. Designed for a couple with three children, the house was partially inspired by the 'Klein Bottle,' a model of a surface developed by German 19th-century mathematician Felix Klein. 'In principal, it's like a doughnut. You can twist and distort it, but it will only change topographically if it's cut. In a sense, there's no beginning or end,' says architect Rob McBride, who worked closely with his partner, interior designer Debbie-Lyn Ryan.

The house is made of compressed cement sheets with a black metal roof, which folds down, in part, to form an external wall. Moonah trees, with gnarled blackened trunks, anchor the house to the steeply sloping site. 'We wanted to evoke a sense of the fibro cement beach shacks in the area. We didn't want to make the house feel too precious,' says McBride. A front door, clad in cork, not only creates an unusual entrance, but also alludes to a cork stuck in a bottle. When the 'cork' is removed, a bright red staircase and walls appear, like liquid solidified around an irregular-shaped lightwell.

克莱因瓶别墅

The entry lobby and laundry are on the ground floor. And although not apparent from the front façade, there is a camouflaged back door providing access to a central courtyard. There are no distinctly separate levels; instead the rooms splay off the staircase around the central void. As you ascend the staircase there is a large, flexible space that functions as a rumpus room and extra room for the children's friends to stay over. Continuing up, there are two additional bedrooms and two bathrooms. To maximize the light, as well as the views over the trees, the open-plan kitchen and living areas are located at the termination of the staircase, as is the main bedroom, a few steps above the living areas. Like the façade, the ceilings fold around the spaces like origami—likewise the fireplace, which appears 'folded' in one corner of the room. The brief to MCR was for a joyous house, one that liberated the senses. And like a great party, where the bottle is left uncorked, the front door is regularly open to extended family and friends. Photography: John Gollings

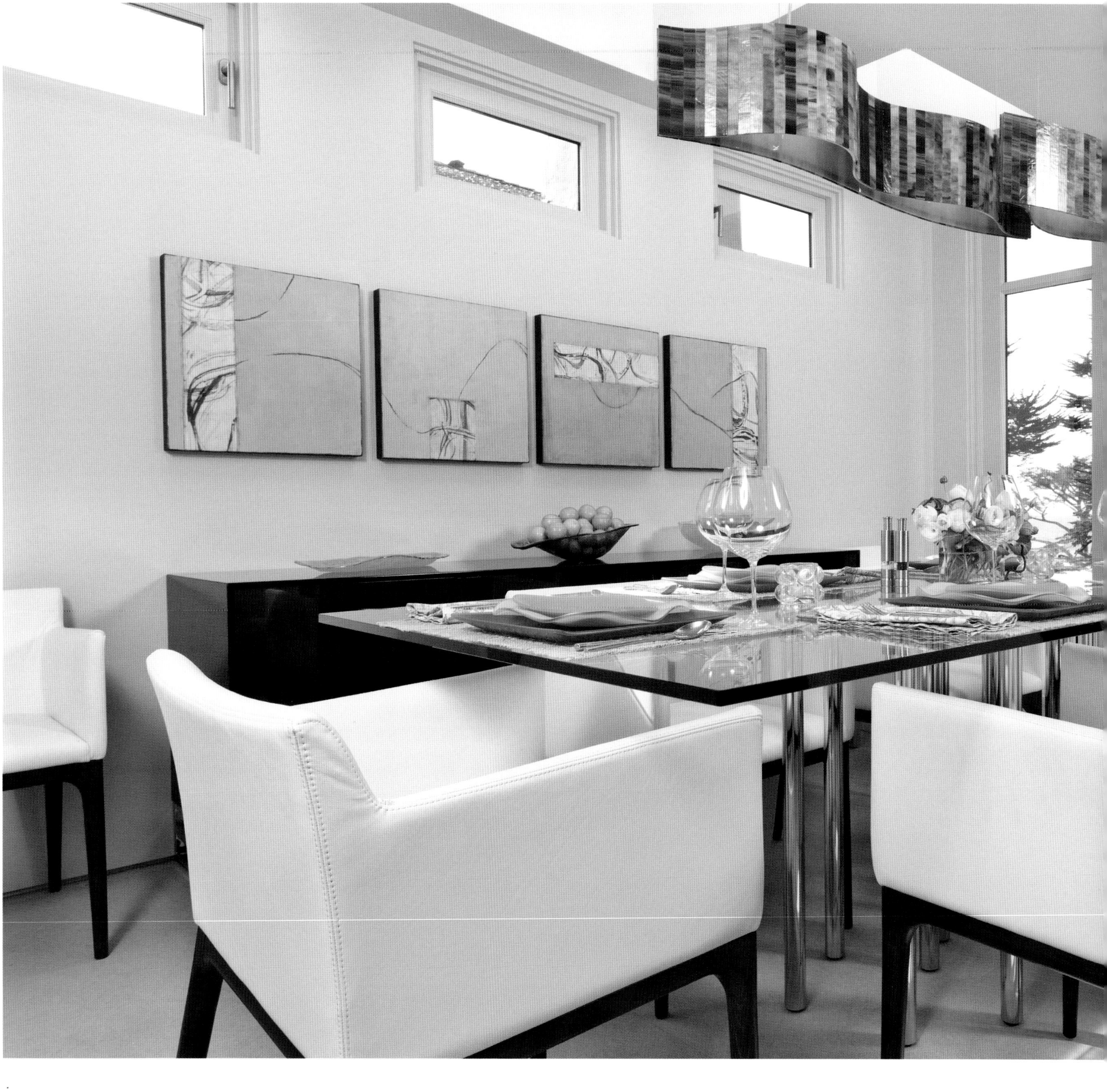

灯笼度假屋

Carmel by the Sea is one of those legendary California coastal towns whose reputation stretches well beyond its central coast region. Its gently rolling cypress and pine-studded hills end at a storybook-worthy crescent beach that forms the western border of the village. Originally an artist colony, with such famous past residents as Ansel Adams, Robinson Jeffers and Robert Louis Stevenson, the town has become a desired weekend retreat for the well to do of San Francisco and Silicon Valley. Building a new house in this well established community usually means tearing down an older, smaller one, and working within strict building codes and deed restrictions.

The **LANTERN HOUSE** was just such a house. The owners, an executive and his wife based in London, wanted a seaside vacation home in Carmel that had a European feel in its fit and finish. They also wanted views of the beach and ocean just a city block away but filtered by neighboring rooftops and mature cypress trees. They hired **Eric Miller Architects**, a firm with strong experience in creating both classic and modern residences in the Monterey Bay Area. Miller's task was to fit a 2700-square-foot (250-square-meter) house on a site that previously held a 1200-square-foot (112-square-meter) cottage, and do it with the same footprint. Miller's solution was to excavate. He designed the house with three levels, one below ground, one on grade and a second story at the rear that sat on grade at the front. Miller did a reverse floor plan, putting the main living spaces on the top floor.

By using a steel frame, Miller was able to create a columnless wall corner on the top floor with sliding glass panel doors that could be completely opened on fair weather days for unimpeded ocean views. Miller connected the three floors with a spectacular glass and steel staircase on the street side of the house. The stair-tower's walls are lined with a buff rusticated limestone and lit by a two-story high window wall with individual windows set in mahogany mullions. The effect is both dramatic and inviting.

Miller's firm did both the landscaping and the interior design work on this project. The master bedroom opens onto a very private rear yard that Miller conceived as a modern painting, complete with Zen-like water feature. At night the house glows like a lantern thorough its many small windows, creating its name: LANTERN HOUSE. Photography: Russell Abraham

LONG DUNE RESIDENCE in Truro, Massachusetts, is perched on a coastal bluff overlooking the Atlantic ocean and was designed by **Hammer Architects** to take advantage of the views of Pamet River and the nearby freshwater pond. The entry side of the house appears very solid with its thick, insulated, wood-clad walls and narrow strip windows enclosing the bathrooms, outdoor showers, stair, and laundry room. Little is revealed until entering the house through a tall glass door that emerges as one approaches the house. Once inside, the living and dining rooms, which occupy the building's center, open to the dramatic water views through a floor to ceiling glass wall that features large sliding doors connecting to a multi-level outdoor deck.

长丘住宅

One wing of the house provides the guest bedrooms, while the other wing, which is rotated forty-five degrees in plan, contains the master bedroom suite. A screened porch with a referential kite shaped roof occupies the intersection of the two geometries, providing views in all directions. Active and passive solar design incorporating photovoltaic panels and deep roof overhangs with ventilating clerestory windows, promote natural cooling and ventilation, while shading the interior in the summer months. Photography: Courtesy of the Hammer Architects

朗埃克别墅

LONGACRES HOUSE is a coastal dwelling on the Burrow Peninsula in North County Dublin, Ireland, that lies on a sand and gravel ridge of salt marshes and dune grasslands. It is among the most scenic and peaceful places anywhere in Ireland that enjoys an abundance of flora and fauna throughout the year. The client's brief proposed that the house's form, scale, and overall visual appearance must evolve from its rural and coastal environment setting. The building consists of a two-story elongated spine off which extend two additional arms at either end, that together form a private sunny courtyard. The principal aim behind this form is to take full advantage of the extraordinary coastal views from most rooms in the house, while capturing the sun's progress throughout the day. All the social spaces enjoy dual aspects so that the coastal panorama and sunlight are present. **Damien Murtagh Architects** adopted an open-plan layout within which voids, frameless glass railings and sliding walls allow playful interaction and communication throughout.

Externally, dry stone walls, brilliant white stucco rendering, patina copper, and cedar cladding blend effortlessly together and into their surroundings. The result is a light-filled home where adults and children alike are in touch constantly with the outside world while sheltered from the elements. A recessed alcove set at the junction of the protective dry stone wall and white rendered elongated spinal core forms the main entrance to the house. Giant slabs of warm travertine stone, set against the brilliant white internal walls, voids to the basement and second floor, and large expanses of glazing create a striking light-filled entrance hall. The open-plan kitchen, dining, and living areas are accessed through a large sliding door. Off the dining area is a raised living space, separated by a cantilevered staircase with a frameless glass rail, encased in white Corian.

Across a bridge that traverses the circulation core are three bedrooms, including the master bedroom, which has a sheltered morning terrace, and a guest bedroom with its own roof terrace. This design outcome responds sensitively yet boldly to both its surroundings and requirements. The design, form, and harmonious synthesis of materials is a modern interpretation of Celtic coastal architecture. Photography: Michael Taylor, Anthony Hopkins

It's difficult to reconcile **LOUTITT BAY VIEWS** with its previous incarnation; a simple, single-story beach shack. The house had been in the one family for several years and a few minor alterations had been made, such as a kitchen upgrade. Owned by a couple with two adult children, the brief to the architects was to maximize views to the north, aligned to the bay. They also wanted to integrate the indoor and outdoor spaces, particularly as the land slopes 15 feet (5 meters) down from the road at the front of the property.

In addition, the clients asked for a separate living area for their children, a place where they could entertain friends. Initial client discussions centered on retaining the original house and extending it. 'We thought about knocking it over completely. But the existing footprint set up the planning parameters. If we started from scratch, we might have been required to build further down the site, thereby reducing the water views,' says architect Peter Woolard, director of **Studio 101 Architects**.

Loutitt湾景别墅

The original freestanding garage, located to one side of the house, was converted into a living area for the two children. Floor-to-ceiling glass doors were inserted in the garage's north wall, leading to a new timber deck. A kitchenette and a bathroom were also included in the new quarters. Substantial changes were made to the original house. The ground and only floor of the original home was completely reworked. It now comprises four bedrooms, including the main bedroom with ensuite, a bathroom, and a separate powder room. The bathrooms are clearly delineated with recycled timber battens, which also form a feature wall for a new staircase.

The new first floor is also clearly delineated. Constructed in timber, the first floor is adorned with Rheinzink panels. 'The Rheinzink captures the color of the landscape. The ribs in the zinc cast their own shadows on the house,' says Woolard, who was also keen to reduce the scale of the second-floor addition. Upstairs, there is a large, open-plan kitchen, living and dining area, flanked to the north by floor-to-ceiling glass doors and windows. And to attract additional light, highlight windows wrap around the living areas. Rather than glimpses of water through tree trunks, there are also panoramic views from numerous vantage points, both inside and from the terraces outside the home. Photography: Trevor Mein

马鲁巴别墅 **MAROUBRA HOUSE** appears to merge with the sea. Designed by **Wright Feldhusen Architects** for a client who is a keen swimmer and wanted to be connected to the water both when doing laps or relaxing inside. While the house enjoys 180-degree views of the water surrounding the Sydney suburb of Maroubra, Australia, there are houses directly in front. But as the house is positioned on a hill, with an approximately 20-foot (6-meter) fall of land from front to back, the vista isn't compromised by neighboring homes.

Relatively modest in size, the 5,382-square-foot (500-square-meter) site was subdivided by the owners. 'Split down the center, two new houses enjoy unimpeded views,' says architect Tim Wright. Designed for a couple with two young children, the house is spread across three levels. Car parking and servicing equipment for the pool are at basement level. The kitchen, living and dining areas, a children's rumpus room, plus a guest bedroom and ensuite are on the first floor. On the second and top level are the main bedroom, dressing area, and ensuite, as well as the children's bedrooms and laundry. As the site slopes toward the street, the laundry is aligned to ground level.

MAROUBRA HOUSE features extensive use of glass, achieving the clients' desire for the house to be as transparent as possible. From the main bedroom or the living areas, water can be enjoyed from most vantage points. Zinc cladding and off-form concrete also make up the range of materials used. Zinc, being a non-ferrous material, resists rust (essential being so close to the sea). 'Our clients wanted a house that was low maintenance. But the house is fairly exposed, so concrete anchors the house to the site,' says Wright. The off-form concrete was poured to appear as though it had grown out of the site, with horizontal bands of concrete forming the base of the house, as well as the walls of the double-height void enclosing the staircase.

One of the most pleasurable activities for the owners is using the pool, either for laps, or just for splashing in. 'You feel as though you're swimming out to sea,' says Wright, who cantilevered the main bedroom approximately 6.6 feet (2 meters) to strengthen the connection between house and water. 'It's only a rocky outcrop below. So the pleasure of the beach comes from gazing out through these windows,' he adds. Photography: Olivia Reeves

The Barrier Islands off the coast of North Carolina represent a natural breakwater for the region. They are mere sand bars less than a kilometer wide in places, but possess some of the prettiest beaches and finest fishing in the Southeast United States. Figure Eight Island, in recent years, has become something of an exclusive retreat for the famous and wealthy of the region. It is here that the owner commissioned well-known Modernist architect **Ligon Flynn** to design and built a large family vacation house for her grown children and grandchildren. The owner wanted a house that could serve as a family compound for three families. Flynn came up with an eco-friendly concept house that exploited the natural ventilation from the ocean.

蛎鹬别墅

OYSTER CATCHER HOUSE sits on the leeward side of the Island but stands on massive 13-foot-high (4-meter-high) posts that allow occasional hurricane-caused storm surges to wash through the ground floor of the house without causing damage. Flynn's concept for the house was to let natural sea breezes rise from the open ground floor and up through the atrium to the third story cupola for ventilation. The main entry to the house is a wire screen-enclosed stair tower that becomes a central atrium for the house. All the living spaces are on the second and third levels, and are arranged around the open-air stair tower that becomes a central atrium. Weatherproof glass walls and doors face the open-air atrium and allow for climate control even though temperatures are often mild. Flynn's unusual design has an open and airy quality that can work well for large family gatherings. Sliding glass doors can open onto both the atrium and external decks, providing flow-through ventilation and sublime ocean vistas.

Jane Ellison, the interior designer, selected a soft palette of aquas, olives, grays, greens and whites that mated well with the surrounding marshlands and seascapes. Her furniture choices are mostly mid-century modern with a mix of the occasional traditional chair or table. The house was designed as a compound for three families and has three master suites under one roof, each with a unique design palette. Flynn fell ill just as construction started and Ellison was charged with the responsibility of doing all of the interior specifications for fixtures and finishes. She had the duel task of fulfilling Flynn's vision, and then adding her own inspiration to it. The end result strongly suggests that she was up for the challenge.

Photography: Kate Carboneau Photography

The **PANEL HOUSE** on Los Angeles' Venice beach sits on a slim lot wedged between other beachfront houses. Its volumes, like spectators at a parade rope line—all shapes and sizes— sit in anticipation of the show before them. The show, of course, is the pedestrian promenade and the beach with the Pacific Ocean as an infinite backdrop. All day long, people stroll, rollerblade or bicycle along the pedestrian path that separates the houses from the beach. They provide a continuous people-watching show that is rich with eccentricity, humor, and old-fashioned family life. Having a picture window on Venice Beach is somewhat like having your own 26-foot-wide (8-meter-wide) flat-screen TV that plays a reality show of L.A. life 24/7.

板式别墅

With the PANEL HOUSE, **David Hertz Architects** has given new meaning to Le Corbusier's famous dictum, 'The house is a machine for living in.' The PANEL HOUSE, embodies a 'machine' that uses recycled materials as a major building component. The wall panels are the same as those used to make industrial refrigeration buildings. They have foam inserts, are skinned with aluminum and painted a dull metallic silver. The panels fit into a muscular steel frame that articulates itself throughout the building. The house stacks up three-plus stories from street level to roof deck, connected by an open steel staircase and open industrial elevator.

Light wells and skylights illuminate the core of the building while expansive window walls open out to the ocean views. The owner is a well-known inventor and entrepreneur. As a consequence, the house contains numerous technologically innovative building conventions not found elsewhere. For example, the entire front window opens vertically, giving a truly unobstructed view of the beach and ocean. The living room fireplace is a small pile of shattered tempered glass with a concealed natural gas source and sits conveniently under the large flat-panel TV suspended on cables. Passersby strolling the promenade are greeted by a compact waterfall that shimmers over a low retaining wall and is topped by a 12-by-20-inch (30-by-50-centimeter) piece of grass the owner has dubbed 'the world's smallest front yard.' The owner, a man with a great sense of humor, relishes both his public persona and access to the ocean. If public display is something of a religion in L.A., Venice Beach is its holy city and the PANEL HOUSE could easily be considered its newest temple. Photography: Russell Abraham

Accessed by a steep driveway, this beachfront house is removed from the localized suburban environment, north of Wellington in New Zealand, confronts the westerly sea view, and is bounded by trees to the north and south.

RAUMATI BEACH HOUSE was conceived as a simple arrangement around a central entrance spine and transverse timber skylight that anchors the center of the house. The sculpted timber form of the central skylight, as well as admitting sun and light in the morning hours, acts as a storage and display zone between the entry, bedroom, and living areas. The kitchen/dining/living space spans the breadth of the west elevation, opening up to the view. In contrast, the bedrooms are more closed, with aspects to north and south via controlled openings and high-level clerestories.

劳马蒂沙滩别墅

The home is designed by **Herriot + Melhuish: Architecture Ltd** with a simple palette of exterior materials including painted board and batten, local river stone, hardwood timber slats and screens, combined with aluminum windows and doors. Some external finishes extend into the interior to accentuate the connection with the outside. Exposed timber framing elements, revealed within some interior walls and ceilings, refer back to a more traditional beach house typology.

Photography: Paul McCredie

Designed by Andrew Simpson of **Andrew Simpson Architects** in collaboration with Owen West and Steve Hatzellis, **SCAPE HOUSE** is not visible from the road. In fact, access to this weekend home is via a 2.5-mile-long (4-kilometer-long) winding dirt road, at the end of which, the ocean reveals itself, as does this unusual black house. 'We saw the house almost as an object, like the granite boulders edging the property,' says Simpson.

Designed for a couple with grown-up children, the house is located on disused farmland on Cape Liptrap, in Victoria, Australia. 'It's a fairly exposed site, with 270-degree views of the ocean,' says Simpson. The brief was not only to capture these views from within the home, but also to have a weekender that was robust and low-maintenance. The clients also wanted something that was as comfortable for large groups of people, as it would be for themselves as a couple.

海角景观别墅

The form of the house, irregular in shape, started with the roof; half skillion-shaped and half a more traditional pitch. And while tiles may be fine for a suburban house, the architects were keen to capture something of the beach culture. A black rubber membrane attached to plywood is evocative of a surfer's wetsuit. 'We used the rubber on the walls and roof to create one continuous form,' Simpson notes. The only contrast to the black is at the point of entry, which features white painted compressed cement sheet walls that continue to the interior. Black steel beams inside the house, set against white walls, also enliven the relatively neutral scheme.

One of the challenges for the architects was 'taming' the Roaring Forties, a wind system that makes sitting outside uncomfortable at certain times of the year. Their solution was to create an internalized 'veranda.' This not only frames ocean views, but also separates the sleeping areas from the living areas. And, in keeping with the owner's brief for flexible accommodation, the veranda also functions as an additional sleeping area for family and friends. 'People tend to be far more relaxed when they go on vacation. They don't necessarily expect ensuite bathrooms with every bedroom,' Simpson points out. Against the ocean backdrop and sky, this house appears as a dark silhouette. 'The form, as well as using black, creates an almost scale-less building. It almost "morphs" depending on where you stand,' says Simpson. Photography: Christine Francis

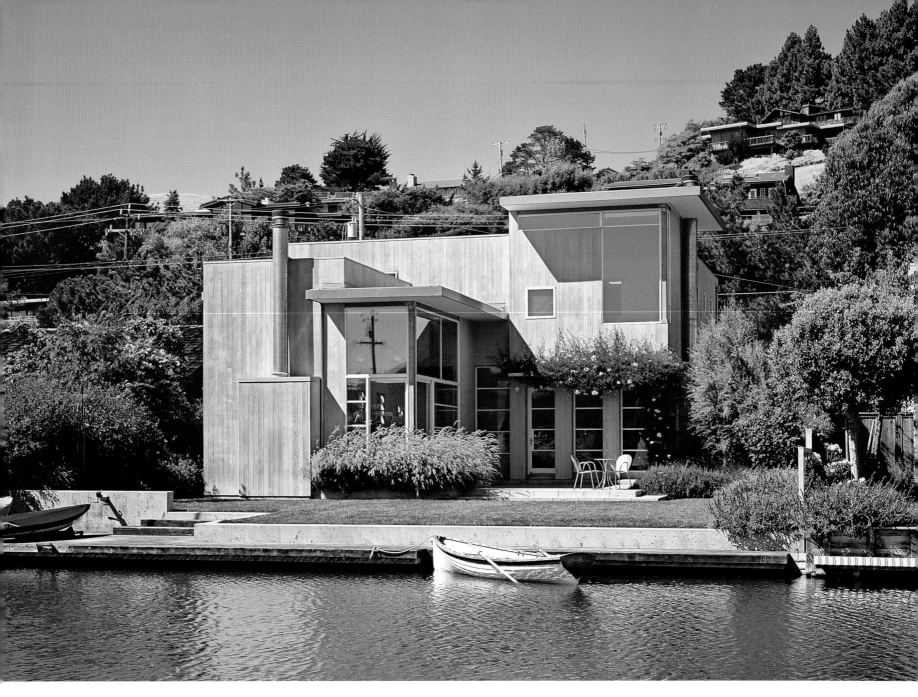

Alexander Seidel designed his new house, in Belevedere Lagoon, in the San Francisco Bay Area, to take advantage of the site by orienting it toward the bay while giving the street a limited view. **SEIDEL HOUSE** is actually not large, but the vertical volumes in the main living spaces create the illusion of a much larger home. The house's red cedar siding has a gray weathered finish that is somewhat evocative of the original fishing shanties that dotted the edge of the bay 50 years before. The use of simple design may be seen as an evolution of what local mid-century modernists like Esherick and Anshen were doing in the 1950s and 1960s. In fact, a number of their houses can be found on the same lagoon.

The elegance of SEIDEL HOUSE is in its simplicity, two volumes broken up by sunshades and a flue. The main living space opens onto a tight patio and garden, which is terminated by a boat dock and the lagoon. The shoreline of San Francisco Bay rolls on for miles, some of it industrial and commercial and much of it undisturbed tidelands, but precious little has been devoted to single-family housing. This Belvedere Lagoon house takes full advantage of its bayside site, giving its owners a modernist waterside oasis enjoyed by few in the intensely urban San Francisco Bay Area.

Photography: Russell Abraham

赛德尔别墅

SUBURBAN BEACH HOUSE, designed by architect **Luigi Rosselli**, enjoys glimpses of the water. Only a short walk to the beach, the original cottage wasn't large enough for a family with three small children and focused mainly on the large house next door. A two-story house, with basement car parking and storage below has replaced the cottage on a reasonably small site of approximately 3,229 square feet (300 square meters), leaving sufficient garden for the children to play in and room for a swimming pool.

The house is constructed in stone, rendered brick, timber, and glass. A sandstone fence anchors the building to the site, which slopes toward the south. 'We wanted to orient the house to the views (south), but also take in the light,' says Rosselli, who was also keen to remove the focus from the house next door. Rosselli created an internal lightwell in the center of the house. A sculptural staircase features a large skylight directly above, as well as a picture window at the top of the stairs. The lightwell ensures less reliance on a northern vista.

郊外海滩别墅

On the ground floor are the kitchen, living and dining areas, which open up to the stairwell-cum-light core. There is also a playroom for the children, as well as a separate study. And on the first floor are four bedrooms, with the main bedroom connected to a large terrace overlooking the pool. The terrace of this bedroom provides a canopy for the outdoor eating area adjacent to the pool. To modulate the light on the second floor, and to reduce the scale of the house, Rosselli used timber for the walls, as well as for the adjustable shutters. To ensure the children are always in sight, Rosselli designed a curvaceous, canteen-style window adjacent to the kitchen. 'The window reminds me of a captain's window in a ship. With three young children, it's not always the weather you need to watch out for,' he remarks. Photography: Richard Glover

Located on the coast of Vestfold, in the southern part of Norway, the **SUMMER HOUSE** replaces an older building that formerly occupied the site. To secure the planning permit, **Jarmund/Vigsnæs AS Arkitekter MNAL** ensured the project was suited to the terrain, in terms of shape, scale, material, and color. The house and terraces are partly built upon existing stone walls. The newly constructed wall sections are made from recovered from blasting at the site.

避暑別墅

The **SUMMER HOUSE**'s low elongated volume is cut in two, creating wind-shielded outdoor areas embraced by the house itself. These divisions also reduce the building's scale and, together with the local variations of the section, ensure the house relates to the surrounding cliff formations. On the outer perimeter of terraces and pool, a glass fence protects against wind while also allowing for unobstructed views. The Kebony wood cladding is treated via a sustainable process to make it more durable against exposure to the salt water.
Photography: Nils Petter Dale

冲浪海滩别墅

SURFCOAST GETAWAY beach house at Lorne, Victoria, Australia was designed for a couple with adult children. The brief was to design a house that would work for the whole family or just for two people, according to architect Ken Charles, who worked closely with fellow **Centrum Architects** director Geoff Lavender.

Overlooking Loutitt Bay, the site is as much about the shoreline as the hilly bush setting, dominated by established eucalypts. As the slope of the site is approximately 30 degrees, the materials used are fairly lightweight: timber and steel frames clad in Ecoply and solid timber straps. Painted a purple-red hue, the Ecoply evokes the color of a sunset at Lorne. Because of the steep slope, almost two thirds of the land couldn't be built on. However, to maximize the building envelope, SURFCOAST GETAWAY is spread over three levels. The top level, also the street level, includes a 'platform' for dual car parking, a study, and an entry vestibule.

Upon arrival, one of the most distinctive features is the sculptural staircase made of timber, glass, painted board, and steel rods. Like the angular posts at the top of the staircase, the timber feature wall is also angled. 'The angles were a response to the angled branches of the trees. They're quite dramatic,' says Charles, who placed a slot window at the top of the staircase and generous glazing below. 'We saw the staircase like a shaft of light coming from the hilltops,' he adds.

On the central level of the house are an open-plan kitchen, living and dining area, all benefiting from panoramic views of the trees and water. There's an impressive view of the water from the ensuite to the main bedroom and providing vistas of the sea from every vantage point was an important part of the brief. Timber features extensively in the interior of the home. Recycled timber was used for the flooring in the dining area, extending to the outdoor terrace. There is also extensive timber joinery, such as a credenza built into the living area. 'It's a fairly robust house. If it's not built-in, then it's extremely solid,' says Charles, referring to the chunky timber table in the dining area.

On the lowest level, designed with separate access, are two additional bedrooms, a bathroom, a billiard room, and gymnasium. There is also a second living area. 'The two floor plates are almost identical. The two areas can be used in their entirety or partially closed down if the owners come down on their own,' says Charles. Photography: Axiom Photography + Design

TAKAPUNA HOUSE is a renovation of a 1980s architect-designed home. The existing architecture had an elegant structure comprising solid banks of bagged brick walls counterbalancing large glazed areas with delicate steel supports. A special feature was the central courtyard space around which the home wraps. However, the configuration of window joinery and poor detailing and finishing had masked the strength of the original concept. The design aim of **Julian Guthrie** was to simplify the building as much as possible and to break down the divisions between the spaces, and between interior and exterior.

塔卡普纳别墅

Opening the house to natural light was fundamental to the design response. The work reveals and responds to the original building rather than seeking to transform or negate it. A primary design decision was to create a new stairwell in the existing central hallway, which allowed a direct spatial and visual link between the two levels of the house. This simple device created a sense of volume and achieved a much stronger flow from the entry down to the courtyard and other lower floor areas. A number of walls were removed from the living areas to create one large informal space that opens up to both the panoramic beach outlook and also the intimacy of the central courtyard.

Minimalist detailing throughout the TAKAPUNA HOUSE gives the various spaces a sense of unity and expansiveness. New joinery openings to both the upper and lower level have maximized the connection to the garden from all rooms of the house. White oiled oak, weathered timber decking, and white rendered walls now give a classic seaside feel to this modernist home. Photography: Patrick Reynolds

Architect **Paul Morgan** was initially captivated by the canopy of tea-trees on this property at Cape Schanck, the southernmost part of Victoria's Mornington Peninsula, Australia. **TEA-TREE HOUSE** sits on a 10,760-square-foot (1,000-square-meter) property, which is depressed, rather than elevated above the street. The area is prone to strong southerly winds so the depression was seen as a positive, rather than a constraint. 'There's something reassuring about being close to the ground in these conditions,' says Morgan, who purchased the site with his sister. The wind factor, as well as the tea-trees, inspired initial designs for the house—aerodynamic in shape and an expression of the strong wind forces in the area. Morgan used materials that wouldn't detract from the tea-trees: glass, plywood cladding stained walnut brown, and composite aluminum panels for the exterior finishes.

茶树别墅

One of the most striking aspects of the open-plan living area is the bulbous water tank. Made of steel and painted white, it collects rainwater from the roof, with the overflow directed to an external tank. While the tank replaces the traditional fireplace as the focal point of the room, it's an important reminder of Australia's drought conditions over the last decade. 'The tank also acts as a structural support, as well as cooling the room,' says Morgan. The tank divides the living area into four distinct parts: the kitchen, living areas, dining area, and a flexible space. Like the organic-shaped tank, the kitchen bench and joinery is fluid in form. Made from laminate and vinyl, the kitchen appears integral rather than separate to the living areas. Handles were concealed on the kitchen joinery to ensure a seamless connection; the unusual cement pavers used for the flooring in the kitchen and living areas are a combination of hexagon and pentagon-shaped tiles.

Unlike many new beach houses, this one is relatively compact—approximately 1,453 square feet (135 square meters) (with an additional 431 square feet (40 square meters) of outdoor decking). 'The house was designed for two families. I'm there about three days of the week. My sister will also be using it,' says Morgan, who sees the working week becoming more flexible. 'There's no longer the strong division between work and leisure there used to be. Cape Schanck is only an hour and a half from Melbourne. Many people spend that amount of time commuting,' he adds.

Photography: Peter Bennetts

里维埃拉别墅

Shubin + Donaldson Architects' designed this house to be environmentally sustainable on many levels—relationship to the site, reuse of foundation, built-in devices, photovoltaic heat capture, radiant floor heating, and the use of natural and sustainable materials. A relatively small house, **THE RIVIERA** has all the elements of an approximately 5,000 or 6,000-square-foot (150 or 180-square-meter) house in a tidy, 3,200-square-foot (975-square-meter) package.

Situated on a ridge, the near-perfect location commands a 270-degree view of the Pacific Ocean below and a view to the Santa Barbara Mountains above. Each room affords great vistas in every direction, as well as stunning natural light throughout the day. The three-level home and garage incorporate an open living-dining area, kitchen, master bedroom, master bath, guest bedroom and bath, powder room, home office, outdoor dining area, outdoor lounge area, and swimming pool/spa.

A monumental feeling is emphasized because the house constantly opens up to the outdoors. The colors of nature are the predominant palette and are a perfectly complementary color scheme. An infinity pool just outside the living room furthers the idea of expanse. A dramatic glass canopy ceremoniously marks the entrance, bisecting the ground-to-roof planes of glass that form sidelights and clerestories. Walls intersect with glass throughout in a play of solidity and transparency.

Floor-to-ceiling bookshelves complement an imposing monolith of mahogany on the living room wall that houses an entertainment center. Set into the wall, and surrounded by floor-to-ceiling glass, it appears as an extension of the outdoors. Doorways in general—even in the limestone-clad bathrooms—are taller than usual and lead the eye upward to be rewarded by either natural light or a beautiful vista. Dark wood floors and softly minimalist furniture are sophisticated and inviting.

A terrace surrounds the house, continuing the indoor/outdoor feeling and accessibility. Bedrooms and bathrooms look out to the ocean. The kitchen faces the hillside, emphasizing the connection with nature. By taking up minimal space, the house also takes up minimal resources. During construction, whatever could be reused was incorporated into the building. Finally, there is a certain efficiency of design in the layout, yet it provides all of the amenities so that the house looks and feels like a five-star private residential club. Making concessions to nature reaps great rewards. Photography: Ciro Coelho

特里格·普安住宅

TRIGG POINT RESIDENCE creates a fluid body of space in synergy with the dynamic coastal context. Space is activated through a sequence of view corridors that engage with distant and close views, which are given meaning through depth in plan and section. The occupant is further engaged through a direct sensory experience that includes touch, smell, sound and sight.

Manually operated devices enable occupants to control comfort and exposure levels responding to the individual's daily and changing desires. One may stand on the edge of the building exposed to the salt, wind and sun or retreat deep into the plan held in the secure arms of the building.

Typically, Perth coastal houses become bunkers of air conditioning and block out curtains. The west sun and southwest winds are continually in conflict with the direction of view, compromising the pleasure of engaging with the exterior. **Iredale Pedersen Hook Architects** introduced an inhabitable double skin space that enables the occupants to control comfort levels in all seasons and on all days.

Fluid space is held momentarily through the manipulation of the ceiling plane and wall planes. Recesses in the ceiling, operable cabinets, doors and walls that disappear or fold define space and allow continual manipulation of the environment.

The house bends and curves to the direction of ocean views while maintaining neighbour view corridors. It launches in to the street announcing its presence and determination to be an active participant in the environment. For Le Corbusier the "house is a machine for living in", for the clients the house is a place for sailing, and for the architects, this house is a place for continual dynamic engagement. Photography: Courtesy of Iredale Pedersen Hook Architects

TRURO RESIDENCE is a modern, high-performance house, situated on a coastal bank 115 feet (35 meters) above the surrounding bay on Cape Cod, was constructed using environmentally sound principles. The design by **ZeroEnergy Design** conserves water use, features native landscaping, and sources nearly 75 percent of its energy from renewable sources, including a roof-mounted solar array.

The two shifting volumes of the house appear to drift within the coastal topography, each form expanding toward the ocean. The glazing asymmetrically wraps and exposes the corner of the interior living space, bringing in light and capturing the majestic ocean view. The tapered roof plane slopes up and out toward the view, drawing the eye outside while the glazing draws the horizon in, inviting the ocean to complete the dynamic fourth wall of the space.

These two primary volumes accommodate extreme fluctuations in seasonal occupancy. The living bar comprises the kitchen, living and dining areas, a large outdoor decking area, and a guest suite—everything needed for a couple's weekend trips. The sleeping bar can accommodate up to 20 people and can be decommissioned to conserve energy when vacant, effectively halving the size of the house. The house features a super-insulated building envelope, durable finishes inside and out, geothermal heating and cooling system, radiant heating, a solar electric system, and fresh air ventilation, which ensures healthy indoor air quality. Photography: Eric Roth

特鲁罗住宅

The clients for **V2V HOUSE** were an entrepreneur and his wife, who bought a tear-down cottage on a small lot in Manhattan Beach. They wanted to build a house that would maximize their allowable building limit and simultaneously sit gracefully on the 'walk-street' it would face. The lot, which was virtually half the size of a typical lot in the sandy section of the city, required **Studio 9one2** to use every nook and cranny efficiently.

Studio 9one2's design approach was to exploit the lot's southwest exposure for both light and views. Behind an attractive patio of concrete planters and native plants facing the walk-street, the house rises up three stories. It features a series of dramatic, glass-railed decks, offset by Brazilian hardwood panels and Jerusalem stonewalls. By using a track-back window system on each level, architect Patrick Killen brought the outside in, creating an illusion of spaciousness. In keeping with the 'less is more' theme, Killen brought the deck tile into the living spaces. This blurs the lines between interior and exterior spaces, allowing the decks to seem larger when the glass wall is slid back fully. Conversely, when the window wall is closed, the floor tiles meet the wood floors and furniture delineates circulation space. Photography: Courtesy of the Studio 9one2

V2V别墅

Killen anchored the stacked deck corner with a spa/pool on the ground level. The 'spool' tucks neatly under the second floor overhang giving the owners a cool, freshwater plunge or hot tub on demand! Bounded on two sides by track-back window walls, the spool and adjacent family room can become one outdoor room when the window walls are opened up. This space fronts on a walk-street that gives a community feeling, while providing additional open space to enjoy the mild Southern California weather.

As with other coastal town homes, Killen placed the public living spaces on the top level using an open plan design to create spaciousness, which also enhances the ocean views. The Jerusalem stone exterior was brought inside to clad the fireplace surround and accents certain interior walls too. Killen used his signature cutout roof overhang on the top floor to help define the building's edge, while adding to its openness.

Studio 9one2 once again shows that working within tight spaces can be an elixir for creativity, economy, and fun.

维多利亚73号

VICTORIA 73 is the result of a dynamic response to capitalize on the site of an existing property. Creating an environment where a young family could enjoy an outdoor lifestyle protected from prevailing winds and enjoy views of the sea and large boulders to the immediate south. The clients were eager to use every possible area of the site, yet ensure privacy in this densely populated part of Bantry Bay in Cape Town, South Africa. The plot's steep tapered shape complicated the design and hampered the construction process. Budget restrictions required a component of the existing building to be retained, causing consequential problems to design, documentation, and execution.

The need to create a family home drove the design of VICTORIA 73, which accommodates the kitchen, living room, and dining room on one level, within a single space. **SAOTA** ensured that these areas enjoy all-day sunlight, frame views of the sea and adjacent rock features, and connect this level to the private bedroom level above. The secondary living area is a dramatic entertainment space, located on the level immediately below the family level. The pool terrace allows for covered and uncovered areas in which to relax. An entertainment lounge accommodates a bar, beside the outdoor braai (barbecue) area, while a dramatic gazebo structure is perched at the western edge of the pool deck—the perfect place for the owners to enjoy the setting sun.

The top floor comprises the main bedroom, two children's bedrooms with ensuites and a small lounge. Guest and staff accommodation, as well as a private library, are located on the first and second stories, below the entertainment level on the third floor. The ground floor accommodates the entry and a five-car garage. A glass lift connects the building vertically, and an external service stair connects the levels externally.

VICTORIA 73 is strongly influenced by the Californian school of 'Case Study' houses and Miesien Planar Designs, demonstrated by the cantilever roof slab separated from the main off-shutter concrete roof soffit by a dramatic clerestory window. In turn, the slab is supported by a marble-clad wall plane, reminiscent of Barcelona Pavilion's stone walls. The finishes are rich and varied in other areas of the house, including timber cladding in many rooms and sumptuously colored mosaic finishes. Interiors are effortlessly casual and sleek. Each piece of furniture was carefully considered to create a successful fusion of 20th-century design pieces with understated, customized items by Antoni Associates.

Visually, VICTORIA 73 is enriched by a number of interesting finishes and features, including textured stone cladding on various internal and external walls. This contrasts with the roughness of the off-shutter concrete soffit to the living room, dining room, and kitchen on the level below the top floor. Timber cladding and richly-colored mosaic finishes are used to provide further texture and interest in other rooms. The client's eclectic art collection plays an integral part in the interior and adds a dynamic background to this residence's spectacular contemporary architecture. Photography: Stefan Antoni

VILLA IN KAIKOH is designed as a vacation house for the owners to invite guests to enjoy the splendid views of the Pacific Ocean. The site is located on a cliff facing the ocean in Atami, Shizuoka Prefecture, a spa town with natural hot springs, 60 miles (100 kilometers) west of Tokyo. Thermal water is provided directly from its source and inhabitants can enjoy the villa's private spa.

The villa is composed of seven layered platforms continuously connected to each other like a folded ribbon, in a gesture corresponding to the convexly irregular surface of the cliff. The garage is on the top floor, which is accessed from the bridge through a gate on the street. There is an entrance inside the garage, a lift that serves all floors, and a staircase to the spa and exterior bath on the terrace. On the main entrance level there are two bedrooms and a laundry with a utility terrace. Downstairs is a fireplace and the living, dining, and kitchen areas—the main public space in the villa. The platform is extended to the exterior terrace, from which one can appreciate the beautiful scenery of the Pacific with an almost 180-degree view.

Throughout the building, there are various sections, creating a range of spaces with a quality of transparency between each one. The master bedroom, bathroom, and two guest rooms are one story down from the main floor, which contains a small gym, a machinery room, and a storage room. Finally, a hobby room is located in the basement, beside a small vegetable garden.

From its foundations, the villa is made of steel based on a reinforced concrete structure up to the main bedroom level. However, there is no structural column or wall on the living floor level, to guarantee unobstructed ocean views. The steel walls and slabs are composed of lattice ribs sandwiched by steel panels. Due to the steep cliff site's difficult terrain, **Satoshi Okada Architects** ensured construction had to follow a logical sequence. First, small prefabricated steel blocks were manufactured offsite and then welded together onsite. These welded steel blocks were suspended by a crane onto each structural element and stacked from bottom to top. Schematically, the bedroom floor slab is suspended from the upper structure to create the garage's solid shell.
Photography: Hiroshi Ueda

KAIKOH度假别墅

鲸鱼海滩别墅

WHALE BEACH HOUSE is a three-level residence designed for a retired couple needing to accommodate their extended family while also providing for their future accessibility requirements. The trapezoidal site is accessed from a private road at the rear and descends 36 feet (11 meters) on the north elevation. Situated at the top of the southern hill overlooking Whale Beach in Sydney, Australia, the **Cullen Feng** design consists of three overlapping, offset rectangular volumes linked via a wide travertine staircase. The entry vestibule has a frameless glass skylight that floods the stair below with natural light.

The magnificent view is anticipated via a horizontal slot in the vestibule and the compression of the walled staircase. Each level of the house opens up to a northern terrace. The upper level consists of a master suite that can be secured independently from the rest of the house; the middle level comprises the living, dining, and kitchen areas, plus ancillary spaces; and the lower level incorporates bedrooms, a media room, and bathrooms. This lower level opens to a deck that features an offset, cantilevered lap pool with a dramatic glass end that juts out over the bouldered landscape.
Photography: Eric Sierins

泽菲罗斯度假别墅

ZEPHYROS VILLA was designed as a vacation house for clients in the UK who eventually left London to live in Cyprus for good. On approach to the house from the garage driveway, visitors face a stone wall and staircase, while the main living area 'hovers' above. The cantilevered mass upstairs provides welcome shade to the lower ground floor, which accommodates a guest room and an artist's studio. It also suggests a playful curiosity, which invites visitors to the main entrance. The stone for the walls on the lower level was sourced locally from the Pomos region, allowing the lower part of the house to blend seamlessly with the surrounding landscape, and enhancing the sense of weightlessness from the hovering white box above.

The main entrance to **ZEPHYROS VILLA** is through a Corten steel door, which leads to a second staircase up to the living areas. Entry here is to framed views of the sky, mountains, and horizon, with the seascape and Pomos harbour below. Thanks to the weather, Cyprus offers an exceptional outdoor lifestyle, which is maximized by the villa's design which includes a patio with multiple 'faces' that, together with a pool, divides the living areas in two. Sliding roofs retract to expose the sky, and wooden screens, when opened, invite in the local landscape.

Facing north, over the small fishing village of Pomos—the architect's focal point—the property's location and orientation create a soothing serenity. The house is proximal to a cliff and surrounding fields of lemon, orange, and pine trees blend with thyme and other aromatic wild plants. Living in this house is about being exposed to the elements but simultaneously being protected—like camping—sheltered yet only a stone's throw from nature.

Koutsoftides Architects wanted to enhance features within the villa to reflect this idea, so that the building and landscape become one—receiving the glorious sun, the western Zephyros wind, the majestic blue sea, and the mysterious silver moon. The house is a reminder of childhood camping trips and enjoying nature, with the aim of embracing all the elements under one permanent 'tent.' Photography: Christos Papantoniou

索引
INDEX OF
ARCHITECTS

索引

Index of Architects

1 + 2 Architecture 62
www.1plus2architecture.blogspot.com.au

Alexander Seidel 158
www.seidelarchitects.com

Andrew Simpson Architects 154
www.asimpson.com.au

Architecture Saville Isaacs 52
www.architecturesavilleisaacs.com.au

BGD Architects 8
www.bgdarchitects.com

BKK Architects 20
www.b-k-k.com.au

Centrum Architects 166
www.centrumarchitects.com.au

Cullen Feng 198
www.cullenfeng.com.au

Damien Murtagh Architects 128
www.dmurtagh.com

Daniel Marshall Architects 102
www.marshall-architect.co.nz

David Hertz Architects 148
www.davidhertzfaia.com

Eric Miller Architects 72, 114
www.ericmillerarchitects.com

Flesher + Foster Architects 98
www.fosterarchs.com

Hammer Architects 122
www.hammerarchitects.com

Helliwell + Smith • Blue Sky Architecture Inc. 96
www.blueskyarchitecture.com

Herriot + Melhuish: Architecture Ltd 152
www.hma.net.nz

Hulena Architects 10
www.hulena.com

Iredale Pedersen Hook Architects 182
www.iredalepedersenhook.com

Jackson Clements Burrows (JCB) Architects 30
www.jcba.com.au

Jarmund/Vignaes AS Architekter MNAL 86, 162
www.jva.no

Javier Artadi 18
www.javierartadi.com

Jorge Graça Costa 82
jorgegracacosta@gmail.com

Julian Guthrie 170
www.julian-guthrie.com

Koutsoftides Architects 200
www.koutsoftides.com

Ligon Flynn 138

Luigi Rosselli 160
www.luigirosselli.com

Luis Mira Architects 28
www.mira-architects.com

McBride Charles Ryan (MCR) Architects 108
www.mcbridecharlesryan.com.au

MYCC Architecture 48
www.mycc.es

Park Miller 36

Paul Morgan 174
www.paulmorganarchitects.com

Phorm Architecture + Design 70
www.phorm.com.au

Robert Swatt 92
www.swattmiers.com

Robles Arquitectos 46
www.roblesarq.com

Rolf Ockert Design 24, 64
www.rodesign.com.au

Saota 192
www.saota.com

Satoshi Okada Architects 196
www.okada-archi.com

Shubin + Donaldson Architects 178
www.shubinanddonaldson.com

Stevens Lawson Architects 100
www.stevenslawson.co.nz

Studio 101 Architects 130
www.studio101.com.au

Studio 9one2 88, 190
www.studio9one2.com

T01 Architecture & Interiors 56
www.t01.com.au

Wilson & Hill Architects 104
www.wilsonandhill.co.nz

Wright Feldhusen Architects 134
www.wrightfeldhusen.com

Zen Architects 14
www.zenarchitects.com

ZeroEnergy Design 188
www.zeroenergy.com

Every effort has been made to trace the original source of copyright material contained in this book.
The publishers would be pleased to hear from copyright holders to rectify any errors or omissions.
The information and illustrations in this publication have been prepared and supplied by the participants.
While all reasonable efforts have been made to ensure accuracy, the publishers do not, under any
circumstances, accept responsibility for errors, omissions and representations express or implied.